SECRETS OF THE SEQUENCE

How Genetics Is Shaping The Future Of Veterinary Care

DR. GORDON ROBERTS

TABLE OF CONTENTS

Understanding genetics, p. 4

Sequencing the human genome, p. 9

The Canine Genome, p. 15

The Cat Genome, p. 19

The rise of DNA testing for pets, p. 21

Inherited diseases and canine genetics, p. 26

Genetics and behaviour, p. 36

Precision Medicine, p. 45

Genetic engineering, p. 51

Gene therapy, p. 58

Cloning, p. 65

References, p 74

FOREWORD

Dear Reader,

Inside the cells of every living thing lies a complete set of genes that form the instructions needed to create that organism. The study of these genes over the years has provided some great insights into how animals and plants grow and reproduce. Genetics has also been able to identify the causes of disease and explain the passing down of hereditary conditions.

We are living in an incredibly exciting age where scientists are engineering genes in order to modify organisms. This could be by replacing a damaged or mutated gene or by inserting a gene from a completely different species.

The potential of this field is mind-blowing – an end to all genetic disease, cows that can produce human breast milk, cloned pets and even designer babies. In the near future we will no longer be asking how far we can go with genetic engineering but instead asking how far should we go.

If this taste of the future leaves you wanting more, sign up to my regular newsletter at futuristvet.com or follow me on Twitter: @futuristvet

Dr. Gordon Roberts
The Futurist Vet
March 2017

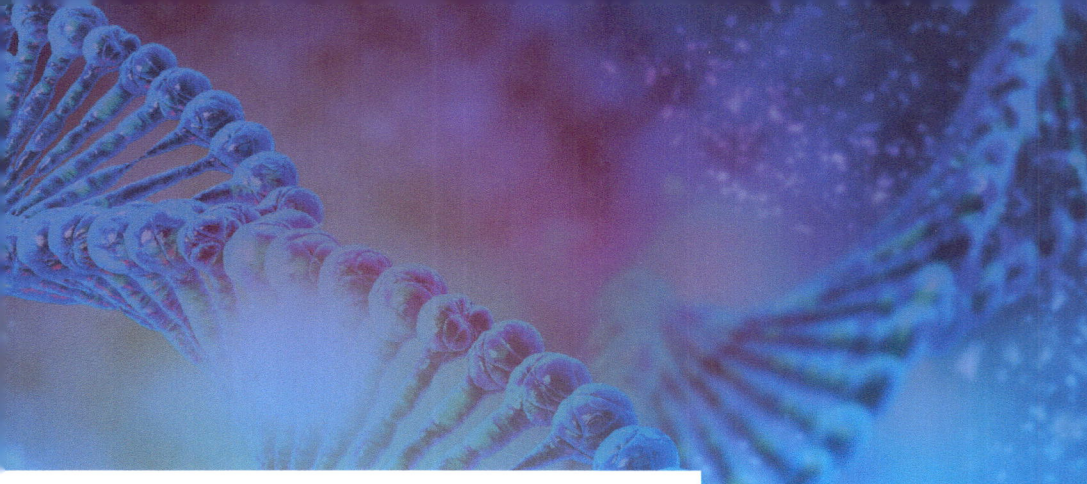

CHAPTER ONE
Understanding Genetics

Since the principles of genetics are the basis for much of what we will discuss in this book, let's begin by briefly discussing what genetics is and what we mean when we use this term.

This basic knowledge will help us to understand the various topics that will come up in later chapters. So, what do we really mean when we talk about genetics in this context? Is this term as obscure and as complex as it sounds? Let's delve a little deeper and find out.

What is Genetics?

Put simply, genetics is the scientific study of how living things inherit certain traits or characteristics from their relatives. These characteristics might be physical in nature, for example eye colour or fur type. Or, they might be behavioural, such as temperament or personality. They may also be medical, such as a predisposition to certain diseases. The ability to inherit these traits applies to all living things, both animal and human.

In this book, we'll be discussing genetics specifically in relation to the animals we keep as pets. Since humans have bred some pets for centuries, we have plenty of fascinating examples of how genetics works and how it has produced the inherent characteristics of today's companion animals. When we talk, for example, about the docile, friendly temperament that a Labrador has, we may well be talking about the selective breeding that has taken place to encourage and reproduce these genetic traits. This is a great example of how genetics, far from being an obscure topic, applies very often to our everyday lives.

Genetic vs environmental factors

Genes are sets of specific instructions for how a cell should grow and develop. Every cell in your body contains genes, from those that make up your skin, to those that are found in your lungs and heart. It's thought that each of your cells contains about 25,000 genes in total. Your genes dictate traits like your eye colour, your hair colour and your blood group. If you've ever wondered why you have the same shaped nose as your father, or the same ears as your mother, you can blame your genes. They are small pieces of biological code, contained within a material called your DNA, which determine all of these characteristics and more.

However, your genes aren't the only influencing factors on how your body grows. Your environment can also be a determining factor. A good example of this is your height. Although you might have the genes to grow tall like your mother or father, if you aren't fed a healthy diet growing up if the environmental factors aren't right, you may not grow so tall after all. A mixture of genetic and environmental factors is at play here.

If we return to the Labrador example, Rover may have all the necessary genes to be a friendly and docile dog, but if he is brought up in a distressing environment he will turn out to be shy and anxious instead, as these environmental factors will override his genetically inherited traits. From these examples, we can see that genes are a deciding factor in how a living thing grows and develops, but we must bear in mind that they aren't the only factor.

Mutated genes

It's also important to bear in mind that not all genes will denote positive or desirable traits. Some genes are defective and these are called mutated or altered genes. With mutated genes, we may inherit certain diseases such as cancers and other medical conditions. This happens when we inherit a specific mutated gene from one or both of our parents. Some examples of conditions that are caused by these genetic mutations include Down's syndrome, muscular dystrophy and cystic fibrosis. Later on in this book, we'll discuss this idea of genetic mutation in relation to our pets, and the particular health conditions that can be passed down in our pet's genes.

Genetic testing

We'll also be discussing the idea of genetic testing, which is the practice of testing our DNA for specific genetic mutations to see if a person or an animal is likely to develop a particular medical condition. This involves taking a sample and analysing the DNA in the cells to see whether these mutations might be present. This can help to diagnose existing conditions, or it can predict the likelihood of developing certain conditions. In the pet world, DNA tests are predominantly used for certain dog breeds and can tell if a dog is predisposed to an inherited condition such as cataracts or cancer. DNA tests are also used in cases where people don't know the breed of their dog. A simple genetic test might uncover that a mixed-breed dog from a rescue shelter is actually a Husky-German Shepherd cross. This kind of information is hugely valuable when it comes to determining the dog's exercise needs, for example.

Looking closely at our genes

Genes can only be viewed through powerful microscopes, and lie within our DNA. If we're going to understand the basics of genetics, it's important to be able to visualise where genes are found within the basic cell structure.

DNA is short for Deoxyribonucleic Acid, a chemical that contains a genetic code for making proteins for living cells. Proteins form the building blocks for all living things. Our bodies are made up of various proteins, from those in our hair and nails (keratin) to those in our blood (haemoglobin). Our genes basically supervise the production of these proteins, dictating how each protein will develop.

So, we know that our genes are contained within our DNA, but where is DNA found? The answer is in special bundles called chromosomes. Chromosomes are found in the nucleus of a cell. Viewed through a microscope, chromosomes appear like two arms which meet at the centre. They contain the threadlike DNA which, in turn, contains our genes.

DNA and its sequencing

The thread-like structure of DNA is actually more like a twisted ladder and we call this ladder the "double helix". These double helix strands are made up of a combination, or sequence, or four different chemicals: adenine (A), thymine (T), cytosine (C) and guanine (G). Your genes are therefore segments of DNA which are made up of a specific sequence of these chemicals. These sequences are instructions for a particular protein to be made in a particular cell, at a particular time.

Chromosomes

So, we've learned that our genes are contained within our DNA, which is in turn contained in chromosomes floating about in the nuclei of our cells. Now let's return to the idea of chromosomes.

In humans, each cell contains 23 pairs of chromosomes – 46 in total.

Why is this important? Well, each of these pairs contains one chromosome inherited from your father, and one chromosome inherited from your mother. This means that there are two copies of each gene in every cell – so, you have two genes for the eye colour trait, for example, and two genes for the trait of blood type.

It's important to note that the number of chromosomes in a human is different to the number of chromosomes in other animals. Dogs, for example, have 39 pairs of chromosomes, giving them a total of 78. A fruit fly has comparatively fewer – just four pairs of chromosomes, giving it a total of eight. The complete set of genes for a living thing is called a genome. The human genome, for example, contains all of the genes necessary to create a human, about 3.2 billion bases of DNA.

CHAPTER TWO
Sequencing The Human Genome

Until recently, scientists knew that humans had a genome made up of sequences of certain proteins (As, Ts, Gs and Cs) but they had no idea what this sequence was. For this reason, groups of scientists came together in 1990 to begin what became known as the Human Genome Project.

The aim of the project was to determine the sequence of the human genome. With three billion base pairs of DNA in the human genome, this was a task of mammoth proportions. It would also have huge implications for mankind; in fact, 3-5% of the overall budget of the project was devoted to researching the ethical, legal and social implications of having this new information at our disposal.

Sequencing the human genome took two decades to complete, involved hundreds of scientists and cost over $3 billion. But how was it done? Let's take a closer look.

The Human Genome Project

The project was funded by the US government and other groups from around the world, with the sequencing itself performed by scientists in 20 research centres and universities in the US, UK, Japan, France, Germany, Canada, and China.

Volunteers provided blood so that their DNA could be sequenced, and their identities remained completely anonymous throughout the process. The finished genome is therefore a mosaic of DNA from different individuals. The entire project was declared to be complete in 2003, and today everyone with access to the internet can view its findings online.

Sequencing the human genome took two decades to complete, involved hundreds of scientists and cost over $3 billion. But how was it done? Let's take a closer look.

Knowing the sequence of the billions of letters (the proteins denoted by As, Ts, Gs, and Cs) that make up your genome is the ultimate goal of genome sequencing. This is no easy feat - the individual letters of DNA in your genes are just eight to ten atoms wide. To make matters even more complicated, these strands of genes are packed tightly together like a ball of yarn. This means that, in order to begin to identify the sequence of letters, scientists need to first break down the long string of DNA into smaller sections.

These smaller sections are then separated and sequenced individually. This is only possible because DNA binds to other DNA if the sequences are the exact opposite of each other. As bind to Ts, and Ts bind to As. Gs bind to Cs, and Cs bind to Gs. If the A-T-G-C- sequence of two DNA pieces are the exact opposites of each other, then they will stick or bind together.

These DNA pieces are extremely small, so scientists need some way to amplify the signal they can read from each of the letters. Most commonly, scientists use enzymes to make thousands of copies of each genome piece. With these thousands of replicas, they can then go about trying to read the sequences. To do this, a special batch of letters is made where each letter has its own distinct colour.

This mixture of special colours is then added to the genome. The specially coloured letters bind to their opposites on the genome, giving us a double stranded piece of DNA with a coloured spot at leach letter. At this stage, a picture is taken of the genome which will show the specific order of the colours. Seeing this sequence of colours allows scientists to read the sequence of the genome.

Later on, a special computer programme is used to "stitch" together all the pieces of DNA and their sequences, so that eventually we get a complete sequence for the genome. Note: this is the most common way that we can sequence a genome, but it isn't the only method that has been used.

Now that the human genome has been sequenced, scientists are working on what the sequences mean and why the sequence of one person's genes might differ from another. The sum of these differences in our DNA is responsible for all sorts of variations in the human race, from what we look like, to how we behave and what we like and dislike.

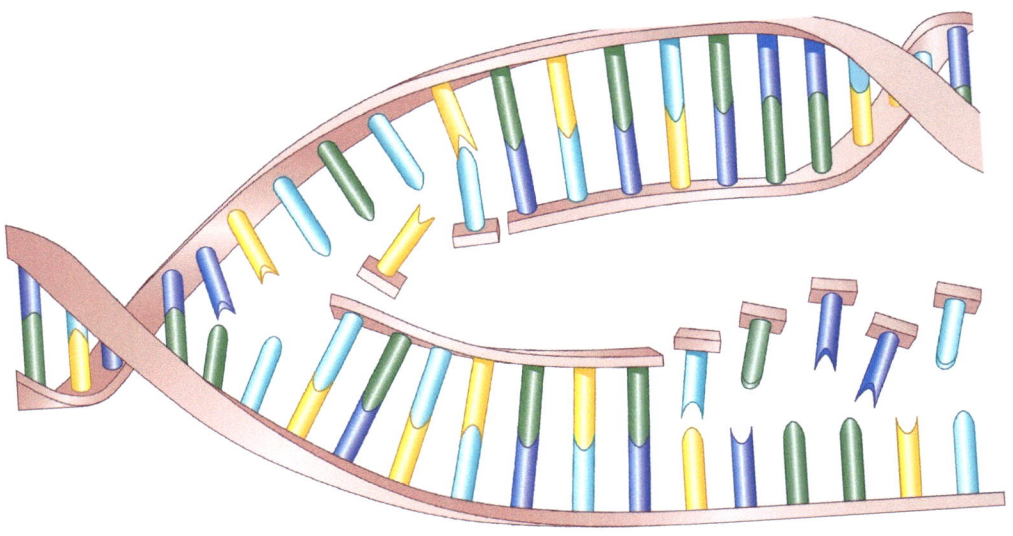

Perhaps most importantly, it dictates whether a person is susceptible to inheriting a particular health condition or disease. Now that they have access to the human genome, scientists are identifying new genes all the time. In 1990 there were less than 10 genes identified by a process called positional cloning. By 1997, this number had grown to over 100.

What does this mean?

The implications of the human genome project are huge and far-reaching. For example, in the coming years we'll be able to detect the genotypes of more and more viruses to determine their proper treatment. We're also able to identify gene mutations linked to particular forms of cancer.

Of course, medicine isn't the only discipline that is affected by the fact that we have now sequenced the human genome. A wide range of industries now have this vital knowledge to draw upon, including agriculture, forensic science, animal husbandry, anthropology and the study of evolution, to name a few.

Having access to the sequenced human genome is opening a vast number of doors for these fields and will bring new discoveries about ourselves as a species.

The 100,000 genomes project

In 2014, the British government announced a £300 million project to sequence 100,000 genomes in England over the following four years. This project would establish the processes required for genome sequencing to become a frequently used tool for the medical field.

This brought with it some good news for cancer research; cancer patients participating in the project would not only have their normal genome sequenced, they would also have the genome of their tumours sequenced at the same time.

The two results would then be compared to try and pinpoint how the patient's normal tissue turned cancerous. Similarly, people with rare diseases who participate in the project would also have their genomes compared to close relatives, in order to pinpoint how their conditions developed.

All of this offers those in the medical field a new way to treat disease - by tracking the changes in a person's DNA from the very beginning, and by finding ways to block these changes from occurring in the first place. The project will provide the much-needed data to begin to research these things.

Genomics

In addition, these advances in the field of "genomics" will allow for better treatment for those with existing conditions. Our knowledge of the genome can help to find the treatment that will work best for an individual, based on their genetics. For example, if a woman's breast cancer is HER2 positive (if it has that specific gene) then we know that Herceptin will be an effective treatment for her.

We can also see whether a particular cancer will respond well to radiotherapy, resulting in fewer radiotherapy sessions for some patients. Genomics can even track infectious disease, helping to identify the source and nature of a particular outbreak, based on the genome of that particular virus or bacteria. Put simply, the scope and potential of genomics, once we have this vital data from projects like the 100,000 Genomes Project, is huge and can lead to earlier diagnosis, more precise diagnosis, new drugs and treatments and even new medical devices.

DNA testing

At the time of writing, it is now possible to have your own genome sequenced for as little as $1,000. This is an astonishing advance – back in 2001, sequencing just one person's genome cost $100 million. Several companies have commercialised the process and are now competing to see who can provide the service the quickest and the cheapest.

To many people, this is an attractive prospect - having your genome sequenced means you will get access to a wealth of information about which conditions you're at risk of developing. Armed with this knowledge, people can make informed decisions about their lifestyles, allowing them to try and mitigate the risks of developing the diseases they are susceptible to through healthier choices. Of course, it is not just our own genomes that we can now sequence, but those of our pets, too. We'll be discussing this later on in this book.

Not everyone is excited about the idea of widespread DNA testing though. Critics say that it will lead to a slippery slope whereby people could potentially be discriminated against on the basis of their DNA. For example, employers could choose not to hire people who have the genes for certain illnesses.

Health insurance companies may also decide not to provide health insurance to these people, leaving them vulnerable due to circumstances beyond their control. Adequate legislation will be required to protect people in these circumstances. It's also important to bear in mind that these issues are just the tip of the iceberg when it comes to uncovering our genetic makeup. Other issues include the emotional effects of discovering this information, and its effects on mental health, as well as family planning issues and the implications of DNA tests on foetuses and babies.

With both positive and negative implications, the only thing we can say for sure is that our knowledge of the human genome, and the rise of DNA testing, will bring huge changes to medicine and to society as a whole. It's also important to bear in mind that these issues are just the tip of the iceberg when it comes to uncovering our genetic makeup.

PET GENOMES
The Canine Genome

In 2005, researchers at the Broad Institute of MIT and Harvard published the canine genome sequence of a female Boxer dog called Tasha. The project, which was part of the National Human Genome Research Institute's (NHGRI) Large-Scale Sequencing Research Network cost around $30,000 and took about a year to complete.

Despite the dramatic differences in the physical appearances of individual dog breeds they are still the same species and, as such, share large segments of their DNA.

A Boxer was chosen to represent the pedigree dog in the Canine Genome Project but its sequence shares the majority of its genetic information with all other breeds.

In fact, researchers found that around 30 per cent of a dog's genetic material accounts for breed variation, including size, shape, behaviour and disease.

Using the sequencing information from Tasha as a genetic 'compass', the research team was able to navigate the genomes of a further ten different breeds, as well as related species including the gray wolf and coyote.

TASHA THE BOXER

The detailed analysis of the canine genome offered for the first time valuable data that could be used not only to understand and improve the health of dogs, but also humans.

The search for disease

Knowledge of the canine genome in combination with the human genome is helping narrow the search for genetic contributors for diseases such as cancer. Many of the cancers that affect dogs are biologically very similar to those affecting humans and any advances made in the treatment of those cancers will benefit both species.

Eric Lander, director of the Broad Institute and Professor at both MIT and Harvard Medical School, said of the Canine Genome Project: "Of the more than 5,500 mammals living today, dogs are arguably the most remarkable. The incredible physical and behavioural diversity of dogs – from Chihuahuas to Great Danes – is encoded in their genomes. It can uniquely help us understand embryonic development, neurobiology, human disease and the basis of evolution."

Humans domesticated dogs 100,000 years ago but selective breeding for desirable traits really started with the Victorians and this is when the breeds as we know them began to be formed. While these breeding practices aimed to preserve and promote certain physical and behavioural characteristics it also led to the passing down of genetic disorders. As a result, many dog breeds are now predisposed to genetic conditions such as heart disease, cancer, blindness, deafness, epilepsy and hip dysplasia.

Helping to read the human genome

Mapping the canine genome has actually made the human genome easier to read. The dog's genome is smaller (19,000 genes compared to 23,000) and due to the human-controlled practice of dog breeding, the gene pools are very small making their 'maps' simpler. Once a dog's map can be read it is then easier to read that of the human by comparing the similarities.

Dr Elaine Ostrander, chief of (NHGRI)'s Cancer Genetics Branch, says, "The leading causes of death in dogs are a variety of cancers. Many of them are very similar biologically to human cancers. Using the dog genome sequence in combination with the human genome sequence will help researchers to narrow their search for many more of the genetic contributors underlying can cer and other major diseases."

Dr Ostrander's team has collected and banked more than 50,000 DNA samples from cheek swabs taken from all varieties of pedigree breeds, having enlisted the help of breed clubs and pet owners.

Pedigree breeds are being used for research purposes because they are genetically isolated, as dictated by the Kennel Club breed standards. By using pedigree dogs that are registered with the American Kennel Club Dr Ostrander is able to obtain genetic data up to ten generations in a canine family, which gives a lot of statistical power when looking at diseases associated with aging.

Dr Ostrander and her team are finding genetic markers on the canine genome much like street names on a map. These signposts point to an area of the genome in which researchers can then look for a gene linked to a specific disease. Before genomes were mapped, finding the right area to isolate was like trying to find a needle in a haystack.

Comparative genomics

Comparative genomics is research that uses predominantly computer-based analysis to compare the complete genomes of difference species in order to pinpoint similarities and differences between them. Comparative genomics is assisting in better understanding the structure and function of human genes in an effort to understand and combat disease.

As DNA sequencing becomes more powerful and less expensive, comparative genomics is now being used in applications from agriculture to zoology as a means of distinguishing between difference species.

Dr Kelly Frazer, a leading expert in genome biology and medicine, says, "It's important to compare the DNA sequences of many species with that of the human genome in order to help decode the human genome sequence itself.

The vast majority of sequences between the dogs and human are different so by looking to see what they have in common it gives us a hint as to what's been conserved because it has function."

Dog leads to narcolepsy discovery

In the late 1990s, Dr Emmanuel Mignot of Stanford Center for Sleep Sciences and Medicine was studying narcolepsy, trying to find the gene responsible for the condition.

Narcolepsy affects one in 2,000 people causing them to leap suddenly from wakefulness to REM sleep. It can often trigger loss of muscle control, sleep paralysis and hallucinations.

Finding the gene responsible was proving difficult so Dr Mignot turned to narcoleptic dogs and looked for the gene in the canine genome. His own Chihuahua, Watson, suffered with the disease, often falling asleep in the middle of a game of fetch when he got particularly excited.

In 1999, Dr Mignot discovered that the condition was being caused by a genetic mutation that destroys hypocretin – a chemical in the brain whose absence causes the disorder.

Being able to find the mutation in the dog genome meant that Dr Mignot could use comparative genomics to locate it in the human genome.

PET GENOMES
The Cat Genome

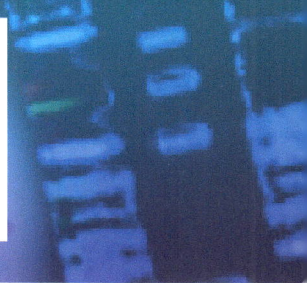

In 2007, The Cat Genome Project based at the National Cancer Institute (Frederick MD) sequenced the first domestic cat genome using a four-year-old Abyssinian cat named Cinnamon. This cat was chosen because its lineage could be traced back several generations.

The aim of the Cat Genome Project was to benefit the health of domestic cats, in addition to serving as a model for human disease. Using the cat genome sequence data, researchers identified several hundred thousand genomic variants, which could be used to determine the genetic bases for common hereditary diseases.

This first attempt at sequencing the cat genome was found to contain significant gaps and errors, and a high-quality version was not published until late 2014.

In January 2015, a group of geneticists, led by Professor Leslie Lyons at the University of Missouri, unveiled the results of sequencing the genomes of 99 domestic cats.

The 99 Lives Cat Genome Sequencing Initiative (affectionately nicknamed Lyons' Den) aimed to achieve a more complete cat genome sequence than previous efforts.

With the rapidly dropping costs of DNA sequencing it has been possible for more work to be done on the cat genome in the past couple of years, although the dog genome is still a long way ahead having been continually improved over time.

Dogs proved to be the better species to be used, due to the small gene pools in pedigree breeds, but cats and dogs both offer insights into human disease, including those associated with old age.

Genomes are like codes and being able to decipher the genetic causes of disease requires a database that describes all the normal DNA variation in cats, as well as the bad mutations.

The 99 Lives Cats Genome Sequencing Initiative relied heavily on fundraising to reach its target number of 99 cats with each genomic sequence costing around $7,000 to complete. The sequencing of the cat genome is not just about understanding genetic diseases but also being able to unravel the mystery of our companion animals' pasts. Razib Khan, evolutionary biologist at UC Davis, wants to use genome sequencing to chart the domestication of cats and to compare different domestic and wild cats around the world.

While the cat genome sequence, like the dog genome sequence before it, is ultimately being used to find ways to improve human health, it is clear that any efforts are going to benefit companion animals too. Prof Lyons was clear about her goal for the 99 Lives project when she said, "I would love to eradicate all genetic disease in cat breeds before we're done."

CHAPTER THREE
The Rise Of DNA Testing For Pets

We have talked a lot about sequencing the human and canine genomes, but what do these advances really mean for our pets? And how do we go about using these findings in our everyday lives?

One of the most widely used applications of our new-found genetic knowledge is the DNA test. Not only are we clamouring to get ourselves tested for certain genetic markers, we're also looking to get our companion animals tested too.

In fact, in the past ten years DNA testing for pets has become a huge industry where several providers are competing in terms of price, accuracy and timescales. In this chapter, we're going to explore the world of DNA testing for pets and what this means for the veterinary world, pet owners and pets alike.

What is DNA testing?

DNA tests involve taking a sample of your pet's cells and sending the sample off to a laboratory where your pet's DNA will be analysed for certain factors in their genetic makeup. There are two main methods when it comes to taking a sample from your pet: a cheek swab and a blood test. Most commercial DNA tests use a cheek swab method that can be done easily at home. After packaging the swab and sending it off to the applicable address, you can expect to wait a few weeks for the results.

Currently, there are two major reasons why a person would want to check their pet's DNA: firstly, to see which breeds make up a pet's ancestry, and secondly, to find out whether that pet has a genetic predisposition to certain diseases.

Breed tests

Have you ever visited a dog shelter and seen the large numbers of dogs that, instead of being purebreds, are "cross breeds"? These dogs have a mixture of different breeds in their bloodline, and because they are rescued dogs, their true parentage is rarely known. One dog, for example, might be 25% Husky and 75% German Shepherd. Another dog might be 50% Labrador and 50% Beagle. Whatever the case, knowing the true breeding background of a dog can be extremely useful when it comes to knowing its energy levels, exercise needs and behavioural traits. It will give you a good idea of how large your dog is going to grow as an adult, and will also give you an idea of which health conditions to keep an eye out for later in life.

With mixed breed dogs, these things are often impossible to find out without the help of a DNA test. In fact, many dog shelters in the US are already using them to help find the most suitable homes for their dogs; for example, some breeds need a home with a big garden, others aren't good with young children. Finding a home to suit the needs of each breed is an important part of the process. It's not only shelter staff who are using these tests - dog owners are also opting for them, hoping to confirm or contest what they have always suspected about their dogs' ancestral roots.

Health tests

Although testing for a dog's breed will tell you a lot about which health conditions it is susceptible to, it won't tell you everything. A more effective way to get this information is to test your dog's DNA for specific gene mutations – that is, markers for particular diseases that have been identified by scientists.

Commercial DNA testing companies are beginning to cater to this need, selling tests that can confirm or deny the presence of specific genetic diseases. Some examples of diseases that DNA tests can identify include: primary lens luxation, progressive retinal atrophy, and degenerative myelopathy.

Different breeds are susceptible to different health conditions, so owners need to first start with this information before deciding which conditions to test for. Boxers, for example, are prone to certain forms of cancer, and Dobermans are prone to bleeding disorders. Knowing about these specific disorders can help owners to be more vigilant, and will mean earlier treatment if they do arise.

Breeding and DNA tests

DNA tests are hugely useful for breeders who want to produce disease-free puppies. They can use these tests to test for the presence of specific gene mutations in prospective parents. This will help them to breed from the right stock, so that they don't carry certain health issues down through the bloodline.

Producing healthy puppies is not only crucial when it comes to breeding show calibre dogs, it is also important for producing healthy family pets that don't cost their owners thousands in vet's bills or cause heartbreak due to premature deaths. In short, when it comes to these DNA tests, everyone wins.

The Kennel Club has a database of inherited disorders that affect each breed, so each breeder can easily find out which conditions to test their dogs' DNA for (a local vet will also be able to provide this information). Most DNA tests can be performed with a simple mouth swab, where a small brush is gently rubbed on the inside of a dog's cheek.

The bristles of the brush collect loose cheek cells which are then sent off to the lab for testing. The UK Kennel Club also has an online resource called Mate Select, where you can search a database of health test results for each Kennel Club registered dog. This is useful when searching for the right dog to parent your puppies.

The rise of commercial pet testing

There are now several companies throughout the world who are offering these DNA tests for pets. The tests available include canine, feline, equine and avian DNA tests. Dog DNA testing is by far one of the most popular and commercialised of these.

Let's take the startup Embark as an example. Embark allows you to learn the breeds in your dog's ancestral history as well as screen your dog for over 160 inherited diseases.

This is a great way to find out all sorts of information about your dog, and most importantly it's a way to find out which inherited conditions could affect your dog throughout its life.

Embark's DNA test tracks over 200,000 genetic markers, giving owners an overview of their dog's risk of genetic disease as well as other inherited traits such as how big they will grow and whether they will shed a lot. This means that dog owners can plan for the future and provide the best possible personalised care for their pets. But DNA tests like Embark's do not come without their problems. firstly, they can be costly.

Generally, the more breeds that a company has in their database, the more expensive the test will be. With 150 breeds to test against, Embark's DNA test comes in at $199.00 which is at the top end of the scale.

Accuracy can also be an issue. Although these tests can identify the majority breeds within a dog's ancestry, if a dog has a large number of breeds within its history, the test might be unable to identify these because there will be insufficient data. However, if the dog has a purebred parent or grandparent, the results will usually be very accurate.

Embark has partnered with Cornell Vet School and in 2016 they published the biggest study of dog genetics to date, with findings from over 10,000 dogs around the world. During the project, researchers successfully identified segments of the canine genome that are associated with shedding, body size and inherited diseases. They plan to collect research data on the dogs whose owners buy their DNA test, and use this information to cross reference aspects like behaviour with a dog's genetics.

Other uses for canine DNA testing

Several city councils around the world have recently adopted doggy DNA testing for a very different purpose: to find and punish those who fail to pick up their dog's poo. These government sponsored schemes will see council workers taking samples of the offending poo and bringing it back to a laboratory for identification. They will then cross reference the DNA found in the poo with a large database of dogs in the area in order to identify the perpetrator. The success of these schemes relies on the ability to test and record the entire canine population of an area or neighbourhood. However, they are catching on fast in the U.S., U.K. and Europe.

CHAPTER FOUR
Inherited Diseases And Canine Genetics

So far in this book, we've learned that the health of our pets is largely at the mercy of two factors: those that are genetic and those that are environmental.

We still know comparatively little about the genetic mutations which cause our four legged friends to fall ill, but we're getting there.

In this chapter, we're going to go through some of the most commonly inherited conditions in dogs, and what we currently know about how these are inherited. This will give us an idea of why the field of genetics is so important to the health and wellbeing of our pets.

Genetic Conditions in Dogs

If you have an interest in dog breeding or genetics, you might already know that the list of genetically inherited conditions in purebred dogs is getting longer and longer as time goes on.

Why is this? Contrary to popular belief, it is not because of poorly manufactured dog foods, environmental toxins or so-called unnecessary vaccines. Those who are biologists or genetic scientists know that the increase in disease is due, in fact, to the prevalence of certain breeding practices.

> **Carol Beuchat writes for the Institute of Canine Biology: "The increasing burden of genetic disease in purebred dogs is a direct and predictable consequence of breeding practices that increase the expression of deleterious alleles harmful mutations."**

In short, the proliferation of these diseases is due to inbreeding, which causes an increase in the expression of gene mutations. In the wild, canines would have had a large pool of other dogs to mate with, causing genetic diversity and leading to a healthy species. However, the domesticated purebred dog is subjected to selective breeding with other dogs that are genetically similar. This is done purposely by breeders who would like to encourage or maintain certain traits in their dogs. Unfortunately, this leads to less genetic diversity and more genetic mutations that lead to harmful diseases.

How does this happen? Put simply, every animal's genome contains many mutations, but because these mutations are usually recessive (i.e. they need to be inherited from both parents in order to take effect) they tend to be passed down from generation to generation without any harmful effects.

The normal copy of the gene is the dominant one and as a result the animal stays healthy. However, if the gene pool is restricted and there are more animals than usual with these recessive mutations, the likelihood of them being passed down by both parents (i.e. becoming what we call homozygous) is increased and this leads to issues.

With two copies of the mutated gene and no normal copies, the resulting offspring could be affected with anything from shorter legs or a different eye colour, to blindness or cancer. We will talk more about how this happens later in this chapter.

The Causes of Genetic Disease

The practice of breeding purebreds began en masse in the 19th century. At this time, breeding dogs for certain desirable traits (such as coat type, size, face shape, etc.) became a very fashionable pursuit. Breeding dogs became a hobby that spread throughout the world, leading to many of the modern dog breeds we see now. This has led to a purebred population of dogs that might look good in the show ring but may also be susceptible to a growing range of inherited diseases because they have been bred from a small gene pool.

As a result, the main threat to our pet's health nowadays is not infectious diseases, poor nutrition or environmental toxins. It is genetic diseases. Some of the causes of genetic disease include:

Genetic drift: the random changes that occur in an animal's genome. This is thought to be due to the chance disappearance of particular genes as individuals die or do not reproduce.

Popular sire syndrome: When a male purebred is seen as a "star" and wins in shows, he will usually then be used to sire (parent) several litters of puppies that year.

This means that a large number of puppies in that generation will carry a copy of his gene mutation, making that mutation (and its corresponding disease) suddenly a lot more common in the breeding population.

This will affect later generations as those puppies go on to breed and the mutation becomes more and more prevalent.

This is a simplified version of what happens, but it's easy to understand how "popular sire syndrome" can lead to serious issues, especially with interbreeding between half-siblings and different generations.

Several real-life examples of this have been found amongst breeders. For example, a high number of Dobermans die from sudden heart failure at an early age. This is the result of a condition called cardiomyopathy which can be traced back to a group of seven popular "sires" in the 1950s (Beuchat, 2013).

Linked traits: It is thought that selective breeding for certain traits can lead to genetic disease because these traits are linked on chromosomes to certain disease-causing genes.

Different Types of Inherited Conditions

The best way to avoid these harmful breeding practices is through DNA testing. This can determine whether a sire (male) or dam (female) should be bred from, or whether they might produce unhealthy puppies.

As we discussed in the previous chapter, this DNA testing is done with a simple cheek swab or blood test and is something that all good breeders should be doing. When it comes to testing a dog's DNA for specific health conditions, there are three types of conditions that can be tested for: autosomal-recessive conditions, autosomal-dominant conditions, and linkage.

Autosomal-Recessive Conditions

For a dog to be affected by an autosomal-recessive condition, there must be two copies of an abnormal or mutated gene (as opposed to just one). As we know from previous chapters, each dog will inherit one copy of a gene from its mother, and one copy from its father. If the health of both the mother and father's genes are known beforehand, then the health of the puppies can be predicted accurately (regarding the condition concerned, at least).

Once each dog has been tested, and the test results have been favourable, they can be used in a responsible breeding programme without the danger of producing unhealthy puppies. Once tested for autosomal-recessive conditions, dogs can be classed as either "clear", "carrier" or "affected".

Clear: The dog carries no copies of the abnormal gene that has been tested for. This means that the dog won't be affected by the health issue in question, and will not pass it on to their puppies. They will only pass on a normal copy of the gene concerned. These dogs can mate with any dogs (whether clear, carrier or affected) without the risk of producing affected puppies.

Carrier: Dogs in this category carry one copy of the normal gene and one copy of the abnormal or mutated gene related to the particular health condition you are testing for. Remember that in autosomal-recessive conditions, the dog will only be affected by the condition if it has two copies of the defective gene. So, in this case, the dog is still healthy.

However, they may either pass on the copy of the normal gene, or the copy of the abnormal gene, to its puppies. These "carrier" dogs can only be mated to "clear" dogs if we are going to avoid the chances of them passing on the health condition to the puppies. In other cases, where a carrier is mated to a carrier, or a carrier is mated to an affected dog, there is a risk of the puppies being affected by this particular condition.

Affected: Dogs in this category are not only carriers of the health condition but are affected by it, too. They possess two copies of the abnormal gene being tested for. These dogs will pass on one copy of the abnormal gene to their puppies if mated. The only dogs they can be mated to without the risk of producing unhealthy puppies are dogs that are "clear". The Kennel Club strongly advises against mating any dogs that could lead to affected puppies. This is classed as irresponsible breeding and should be avoided at all costs.

However, breeding from carriers should not be completely avoided. This is because breeding from only "clear" dogs can lead to less genetic diversity and more inbreeding, which in turn can lead to new inherited diseases occurring.

So, carriers are an important part of maintaining genetic diversity in any breeding programme, if they are bred from correctly (i.e. only with clear counterparts) and responsibly.

It's important to note that clear dogs are only ever "clear" for the particular health condition they have been tested for. This means that they could well carry other undetected gene mutations which could be passed down to their puppies. As such, the Kennel Club cautions that no dog is every really risk-free when it comes to genetically inherited diseases.

However, through responsible breeding programmes and thorough use of DNA testing, those risks can be lessened where possible.

Autosomal-Dominant Conditions

In contrast to autosomal-recessive conditions (where two copies need to be inherited), a dog only has to inherit one copy of an abnormal gene to be affected by an autosomal-dominant condition. Dogs that have been had DNA tests for autosomal-dominant conditions fall into three categories: "clear", "heterozygous affected", or "homozygous affected".

Clear: the dog has no copies of the abnormal gene related to the health condition in question. This means that the dog is not affected with the disorder, and won't pass it on to its puppies.

Heterozygous affected: These dogs have inherited one copy of the abnormal gene from their parents, and this means they will be clinically affected by the health condition. They may (or may not) pass on one copy of the abnormal gene to their puppies.

Homozygous affected: These dogs possess two copies of the gene in question, meaning they are clinically affected by the health condition, and they will also definitely pass on one abnormal copy of the gene to their puppies.

Again, matings which could potentially produce affected puppies should be avoided at all costs. If any puppies turn out to be clinically affected, they must receive prompt veterinary treatment.

DNA Linkage

Sometimes scientists can't find the exact gene that is known to cause a particular health condition. However, they might be able to pinpoint approximately where in a dog's genome it is located. Certain groups of genes are often inherited together because they are near each other on the same chromosome.

So, while it might not be possible to identify the exact gene that causes a condition, scientists might sometimes be able to identify particular sections of DNA that are linked to the gene in question.

Knowing or identifying these linked genetic markers can help breeders to know the genetic status of their dogs. Of course, this type of DNA test isn't quite as accurate as the other types we have mentioned here, because they merely rely on the link between the genetic marker and the disease.

However, they can still be very accurate and you can often get an estimation of their accuracy from the laboratory carrying out the test. These linkage tests are used for both autosomal-recessive and autosomal-dominant conditions.

The Most Common Genetic Diseases

Let's take a quick walk through of some of the most commonly inherited conditions in dogs. You may have heard of some of these conditions before, especially if you're a breeder or an animal professional.

Hip dysplasia: Hip dysplasia is perhaps the most commonly inherited condition when it comes to dogs, affecting several well-known breeds including German Shepherds, Rottweilers, Bulldogs, Great Danes, Saint Bernards, Neopolitan Mastiffs, and Retrievers.

In this condition, the ball and socket joint of the hip is malformed and fits badly together, causing lameness in the dog. Depending on the extent of the malformation, affected dogs will experience considerable arthritic pain and may have to undergo corrective surgery if necessary.

Although hip dysplasia is a genetically inherited condition, scientists haven't identified a single gene that causes it. Instead, this condition is seen as "multifactorial", meaning there are several genes responsible. Environmental factors such as weight gain and diet can make the issue worse.

Heart disease: This condition represents a range of heart defects and issues that can be inherited in dogs. One example affecting Cavalier King Charles Spaniels and Dachshunds is myxomatous valve disease, where pressure develops in the chambers of the heart.

Another condition is dilated cardiomyopathy where there is abnormal heart muscle and a weakened or dilated heart. Doberman Pinschers, Great Danes and Boxers are susceptible to this condition.

Epilepsy: Much like epilepsy in humans, this condition occurs when the cells in the brain get over-excited to a certain degree. This leads to a seizure where the dog stiffens and falls to the ground, paddles its arms and legs and may lose control of its bladder. Seizures can be upsetting for owners, and often there is no known cause.

Urinary bladder stones: "Stones" are a build-up of certain substances within the urinary tract. These can be painful and uncomfortable for the dog, and often there are signs such as straining to urinate, frequent urination, loss of bladder control or blood in the urine. Breeds that are prone to bladder stones include Newfoundlands, Bichon Frises, Miniature Schnauzers and Dalmatians.

Degenerative Myelopathy: This condition is characterised by the slow deterioration of nerve fibres within the spinal cord. This results in a failure of nerve signal transmission which leads to weakness, wobbling, dragging of the hind feet and eventually paralysis. These symptoms usually develop in dogs that are older in age. Breeds which are prone to this condition include the Bernese Mountain Dog, Pug, German Shepherd and Boxer, among others.

What you can do

All of this might sound pretty daunting to dog owners and breeders alike. After all, the conditions mentioned above are just the tip of the iceberg when it comes to the long list of genetically inherited diseases in dogs. But there is hope.

If you're a dog owner who is hoping to get a new dog, and you're absolutely sure you want a purebred, the best way to avoid getting an unhealthy puppy is to simply find a reputable breeder.

That means someone who:
- Screens all of their breeding stock for the major diseases that the breed is susceptible to
- Can provide evidence of this, in the form of certificates and relevant documents
- Demonstrates a solid understanding of the basic genetic principles we have discussed here
- Breeds for the love of the dogs themselves, rather than breeding simply for profit
- Allows you to meet and handle the puppies' parents at the same time as meeting the puppies

If you're a breeder, the best thing you can do is to educate yourself about genetics through reading books like this, and test your breeding stock as much as possible. Avoid excessive breeding from a single male and you'll also avoid the genetic pitfalls of popular sire syndrome and its disastrous effects on the population of your chosen breed.

CHAPTER FIVE
Genetics and Behaviour

In the previous chapter, we discussed inherited diseases in canines, but did you know that dogs can also inherit certain behavioural traits too?

In this chapter, we'll discuss the nature of genetically inherited behaviour and what the scientific community is doing to learn more about it. This will show us just how important the field of genetics is to every aspect of our pets' lives.

Breeds and Behaviour

The UK Kennel Club has recognised 213 distinct dog breeds, each with its own set of inherited characteristics. These include size - consider the tiny Chihuahua in comparison to the enormous Irish Wolf Hound - and snout shape, whether long and pointed like the Collie or flattened like the Pug.

However, not all of these traits are visible; some are behavioural and can only be recognised through observation. The majority of these behavioural traits were created purposely, by selective breeding.

Here are two examples: herding breeds like the Collie have been bred to stalk and herd livestock, and guarding breeds like the Kuvasz have been bred to guard and watch over livestock without chasing and herding them in any way.

Other behaviours that have been bred into dogs include:

Tracking: Scent hounds such as Beagles, Bassets, Foxhounds and Coonhounds have been bred to track the scents of small animals and birds

Retrieving: The Labrador and the Golden Retriever, with their soft mouths, have been bred to fetch and retrieve animals that have been shot in the fields.

Running: Sighthounds such as Afghan Hounds and Greyhounds have been bred to use their vision to track prey, and chase it with the aid of their fast running abilities.

The changing role of dogs in society

Dogs have undergone some radical changes since they were first domesticated. Although domesticated dogs display some of the behaviours of their wolf ancestors, their behaviour has also been hugely altered by artificial breeding processes. As we discussed in the previous chapter, breeding suddenly became fashionable in the 19th century and at this time, people began to keep breeding records that traced the lineage of their dogs. Breeds became narrowly defined, with particular breeding standards (attributes that the breed type was preferred to possess). Dogs became popular not for their attributes as working dogs (like the herding, hunting and retrieving breeds mentioned above) but for their looks, and they began to be exhibited at shows.

The result is that the majority of today's domesticated dogs are now merely companion dogs. This means that many of them need a much higher tolerance than their predecessors for inactivity, social isolation and unstimulating environments (van Rooy et al, 2014). Many of the behaviours that were once useful in the field and on the farm are now obsolete, and sadly lots of our dogs are now having to endure long bouts of solitude and confinement. This is undoubtedly leading to a rise in behavioural quirks and issues.

Dogs as candidates for behavioural research

The extensive breeding that dogs have undergone at the hands of humans makes them ideally suited to being the subjects of genetic research. Between different breeds, there is large genetic variation, and within the confines of the breeds themselves, there is relatively little variation.

As Karen B. London, an animal behaviourist writing for The Bark magazine, puts it:
"The dog has accidentally become the greatest genetic experiment in human history, and is thus the perfect system in which to address many genetic questions."

Despite this, identifying the genes responsible for specific behaviours has so far been problematic. This is because, much like complex diseases such as Hypothyroidism, canine behaviours involve both environmental and genetic factors, and usually multiple genetic components are at play (van Rooy et al, 2014).

After the canine genome was sequenced in 2005, there was a proliferation of the availability of genetic tests. However, the majority of these are for inherited diseases and coat characteristics. To date, there is very little research on behavioural traits themselves, despite the fact that canine research would be very useful for the research of human psychiatric disorders.

Why researching behavioural genetics is complicated

Researching behaviour isn't just problematic because of the complex genetic factors at play; it is also complicated by dogs' interaction with their environment. Behaviours should always be viewed in the context of the dog's environment and its experiences whilst growing and developing. Some behaviour is learned and the dog will have decided to repeat it because it has been successful in the past.

Other behaviours might be the result of experiences the dog might have had as a puppy, for example dogs that have been mistreated early on in life are more likely to exhibit behavioural issues such as aggression and fear later on. Nutritional deficits can also affect how a dog's brain develops and how it behaves as an adult. Lastly, it has also been found that the behaviour of the mother can also influence her offspring.

It is thought that genetics can influence canine behaviour in three major ways:
- How information about potential threats is detected and interpreted
- How memories of past experiences are used
- How the metabolism of certain neurotransmitters is altered

To further complicate the study of behaviour, it should be remembered that certain behaviours are strongly linked to each other, being what we would call "co-morbid" (meaning the symptoms of each disorder often present at the same time). This is especially prevalent with anxiety disorders such as separation anxiety and noise phobias.

Lastly, it is also important to note that categorising or "phenotyping" behaviours can also be difficult. Since there are no blood tests or agreed characteristics for most behavioural conditions, it is crucial that any assessments of behaviour remain as objective as possible and that any criteria that overlap are kept in mind.

Current Behavioural Studies

Despite these obstacles, scientists are trying their best to address our knowledge gaps in the areas of behaviour and genetics. There are several large-scale studies happening which, it is hoped, will shed new light on canine behaviour and its genetic components.

The Dog Project

The University of California is currently heading up a research project called the Canine Behavioural Genetics Project (also known as The Dog Project). Its aim is to explore the relationship between genes and behaviour in canines. The research will also examine genetic diversity in dogs, both between breeds and within them.

The implications for this research are far-reaching. For example, its findings could lead to changes in dog breeding practices, and could also change how we evaluate dog behaviour in the future. It could even lead us to insights on how and when dogs were first domesticated.

Taking an ethical approach, the project will not involve breeding dogs with behavioural problems, but will focus instead on the genes of existing dogs. Included in the project are:
- Dogs who suffer from disorders such as panic, fear, anxiety, obsessive-compulsive behaviours or aggression,

- Close relatives of these dogs who are unaffected by these disorders
- Dogs who don't have any behavioural issues and who don't have any relatives with behavioural issues

Participants in the study are being asked to take a cheek swab from their dogs and fill out a very detailed 25-page questionnaire on their dog's behaviour. The advantage here is that dogs are being studied in their own home environments where the behaviours naturally present themselves, rather than being studied in labs where confinement would itself produce behavioural issues, thus skewing the data.

In terms of the research, two main approaches are being used. Firstly, the project is looking at genetic variations within purebred dogs. Since dogs of the same breed have very similar genes, it is much easier to isolate a particular gene variation in these dogs, especially if some of those dogs exhibit the same behavioural issues. It is also much easier to pinpoint these genes within families of dogs where some offspring have been affected by the behavioural issue, and some have not.

The second approach used by the project is to study the genes of dogs that are not related, either within breeds or across different breeds. These dogs are either affected by a behavioural issue or not affected, but if researchers can find particular points at which their DNA is similar, especially when it comes to abnormal behaviours, then it will be of particular interest to the study.

Dognition and Embark's study

Two organisations have recently joined forces to undertake what they hope will be the world's largest study on canine behavioural genetics. Dognition (a company that offers cognitive tests for dogs) and Embark (which supplies DNA testing for pets) have formed a partnership that will give dog owners unique insights on their behaviour.

The study will be led by Dr Adam Boyko, an assistant professor in veterinary medicine at Cornell University and Brian Hare, an associate professor in Cognitive Neuroscience at Duke University. It aims to uncover:
- The origins of certain behavioural traits
- The most effective training methods to use with dogs at different stages of their development
- The extent to which environment and breed affect a dog's behaviour
- The rate of cognitive decline a dog experiences based on its genes and breed

The study is hoping to test over 5,000 dogs in total. However, participants must first purchase both Embark's and Dognition's services before they can take part.

Darwin's Dogs

Darwin's Dogs is another project aimed at uncovering the links between dog behaviour and genetics.

Based at the University of Massachusetts Medical School, the study is headed up by Elinor Karlsson, a professor of bioinformatics and integrative biology.

The scheme is being fully funded by the University, and so far over 11,000 participants have signed up. To take part, dog owners must take a swab sample of their dog's cheek cells and answer 120 questions about their dog's behaviour.

The study focuses on three key areas:
- Established mental disorders such as anxiety
- Human-selected behaviours such as retrieving
- Behavioural "quirks" such as head tilting and eating grass

The research will focus on studying how the dog genome differs from that of the ancient wolf, giving insights into how genetics shapes the behaviour of our four legged friends. This, it is hoped, will shed new light on psychiatric and neurological diseases in both dogs and people.

"We know that psychiatric diseases in humans and dogs have a big genetic component. They're very heritable," Karlsson said, speaking to the Washington Post. "We're trying to understand in much more detail what exactly are the pathways of the brain that are involved in these diseases."

Progress in Genetic Research

As the studies above show, we are currently attempting to better understand the link between genetics and behaviour, both in our canine companions and in ourselves. The good news is that progress is already underway and new discoveries are being made all the time.

Here is an exciting example. In 2016 a study by Linköping University in Sweden discovered that the ability of dogs to interact with humans is strongly linked to their genetics. When they were first domesticated, dogs adapted to life among humans by modifying their behaviour in some unique ways.

This new behaviour helped them to interact and communicate with humans, setting them apart from their wild ancestors, wolves. The study found some strong evidence for this behaviour which points to the ways in which dogs differ from wolves.

More specifically, when faced with a difficult task, the study found that most dogs will respond by seeking contact with a human, whereas wolves will attempt to solve the problem themselves.

"Our findings are the first to reveal genes that can have caused the extreme change in social behaviour, which has occurred in dogs since they were domesticated," Per Jensen, professor of ethology and leader of the research group, told Science Daily. The study focused on almost 500 Beagles with similar early experiences of humans. It presented the Beagles with a challenging task (opening a tight lid to obtain a treat) and recorded their reactions on video to observe their willingness to seek contact with a human in the room.

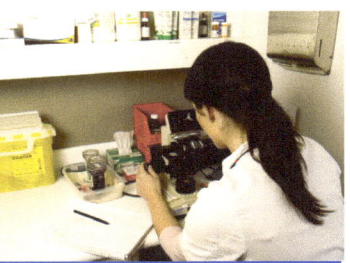

At the same time, scientists studied the DNA of these dogs using GWAS, a method also known as a genome-wide association study. This method is useful in finding out if a particular genetic variant occurs often in individuals who possess a certain trait (in this case, the behaviour of seeking contact with a human).

The study found that the Beagles who did seek human help with their tasks had certain genetic variants in common. In this case, there were five different genes that proved to be of interest to the researchers.

These five genes have also been linked to human social disorders such as autism. The next step in the research is to see if these associations can also be found in other dog breeds besides Beagles. If this is the case, scientists may be able to use this data to better understand social disorders in humans.

CHAPTER SIX
Precision Medicine

As we learn more about the genes of our pets, new veterinary treatments are emerging that will improve the lives of our furry friends.

Precision medicine is one of these treatments. Already making waves in the field of human medicine, it is also set to change how we treat animals in the near future.

In this chapter, we're going to explore the idea of precision medicine and what it could mean for our pets going forward.

What is precision medicine?

In human medical care, precision medicine is a new approach to treating and preventing disease that is tailored to each person's needs. In deciding the right course of treatment, it takes into account the variations in their genetics, environment and lifestyle. As opposed to a "one size fits all" approach, this tailored method allows doctors to more accurately predict which treatments (and prevention strategies) will work for a certain group of people.

The term "precision medicine" is fairly new, but this is a concept that has been around for quite some time. For example, when a person receives a blood transfusion, they aren't given just any blood - they will have their blood type matched to a donor's in order to reduce the risk of complications. This is a basic example, but it's important to know that this type of precision medicine is currently very limited. It is hoped that this approach will, in the near future, be adopted more widely and will soon be found in many different areas of health care.

The term "precision medicine" often overlaps with the term "personalised medicine" and the two are very similar concepts. However, "personalised medicine" is an older term that became outdated. It was thought that the term "personalised" could be misinterpreted to mean treatment that was tailored to individuals. In reality, the idea of precision medicine is not that it is tailored to the individual as such, but that it is tailored to certain groups of people with similar needs.

According to the National Research Council:
"Precision Medicine refers to the tailoring of medical treatment to the individual characteristics of each patient. It does not literally mean the creation of drugs or medical devices that are unique to a patient, but rather the ability to classify individuals into sub-populations that differ in their susceptibility to a particular disease, in the biology and/or prognosis of those diseases they may develop, or in their response to a specific treatment."

Barack Obama made the concept of precision medicine more widely known in 2015 when he announced the Precision Medicine Initiative, a new research initiative where $216 million in funding would go towards expanding precision medicine. One of the aims of the research is to find new, more effective treatments for various forms of cancer, based on new knowledge in the area of genetics.

Pharmacogenomics

Pharmacogenomics is an example of an emerging practice in precision medicine. Pharmacogenomics is the study of how our genes affect our response to certain drugs. The field is relatively new, and combines the study of drugs (pharmacology) with the study of genes and their functions (genomics) in order to develop safe, effective medications that will be tailored to a person's genetic makeup.

A lot of the drugs we use nowadays are "one size fits all" - they are prescribed to everyone with a particular disease, yet in reality they don't work the same way for everyone. Different people will benefit from a medication and some people may not even benefit at all. Some will also experience harmful side effects.

Researchers are trying to change this. With the knowledge gained from the Human Genome Project, they are trying to learn how differences in the genes we inherit might affect how the body responds to medications.

It is hoped that we will be able to use our knowledge of these genetic differences to not only prevent adverse drug reactions, but to know whether a drug will be effective for a particular person. In the future, it is thought that our knowledge of pharmacogenomics will lead to the development of specially tailored drugs for conditions like cancer, cardiovascular disease, asthma, HIV/AIDS, and Alzheimer disease.

Precision medicine in veterinary care

Precision medicine is already finding its way into veterinary care, and will one day be the norm when it comes to treating our pets. Here is a great example: in a recent article for Genome magazine, Kendall K Morgan writes about Hannibal, the Bernese Mountain Dog who had a confirmed case of lymphoma.

If left untreated, dogs with this form of cancer rarely live longer than a few months. Hannibal's owners had heard of an expert called Matthew Breen, who was a professor of Comparative Oncology Genetics at North Carolina State University's College of Veterinary Medicine. So, they booked an appointment to see him.

It turned out that Breen had developed a special test for dogs with lymphoma. The test, by determining the presence of certain genetic factors, could help to predict how long a dog would respond to standard chemotherapy. Hannibal's owners opted for the test, and it showed that he had a 90% chance of still being in remission after seven months. So, they opted for treatment. This is an example of precision medicine already at work in the field of veterinary care.

In Hannibal's case, he received treatment based on certain markers in his genes. This resulted in him getting the right treatment and gave him more time with his family.

Matthew Breen is hoping that his genetic test approach could one day be adapted to human lymphoma treatment. Over the course of his career, he has observed that human and dog cancers share many similarities, and even share the same kinds of chromosomal arrangements in many cases. In this way, he says that human and dog health are "opposite sides of the same coin".

Using gene tests

Precision medicine is still a work in progress when it comes to canine medicine. However, there is a surprising amount of light at the end of the tunnel. Many companies and veterinary schools are already offering a range of genetic tests for everything from inherited genetic disorders to breed identification.

One example, which is perfect for the application of precision medicine, is a test for a mutation in a gene called MDR1 (which stands for multi-drug resistance gene).

A mutation in this gene is linked to a range of adverse drug reactions in dogs, including to the heartworm drug ivermectin, certain chemotherapy drugs and even the over-the-counter drug Imodium. The gene mutation can cause mild to severe drug reactions, depending on whether there are one or two copies of it found in the genome.

The condition most commonly affects the herding breeds like Australian Shepherds and Collies. Having a test for this particular gene mutation means veterinarians can know in advance whether to expect reactions from certain medications and whether to try and prescribe an alternative drug instead. This tailored approach is at the heart of the idea of precision medicine.

The future for precision medicine

The future for precision medicine in veterinary care looks bright, not least because many of its approaches can be used to inform human medicine also. Some recent medical and scientific advances mean that the time is right for precision medicine to become more popular and for veterinary medicine to take on a bigger role in its evolution. These advances include:
- New sequencing technologies, which have become dramatically cheaper and easier in recent years
- Electronic medical records and digital data, which are rapidly becoming more widespread

- The emergence of new mobile devices to monitor and track the health of individuals
- The arrival of sophisticated bioinformatics tools which can interpret large pieces of data

In addition to all of this, today's patients are more willing than ever to take part in medical studies which will help to advance our knowledge in these areas. The human owners of animal patients are also keen to take part in these studies and this availability of test subjects is proving to be a huge advantage to the veterinary field.

Precision medicine has the potential to be far-reaching, and will inform many other areas of medical care, for example disease susceptibility, pharmacogenomics, disease surveillance, risk of surgical complications, and development of drug resistance.

Above all, it will provide the basis for more effective decision making, both in human and in animal care, and will lead to more successful treatments overall. We can only assume that, once precision medicine becomes the norm for human patients, we will naturally work towards using the same principles and techniques to benefit our pets.

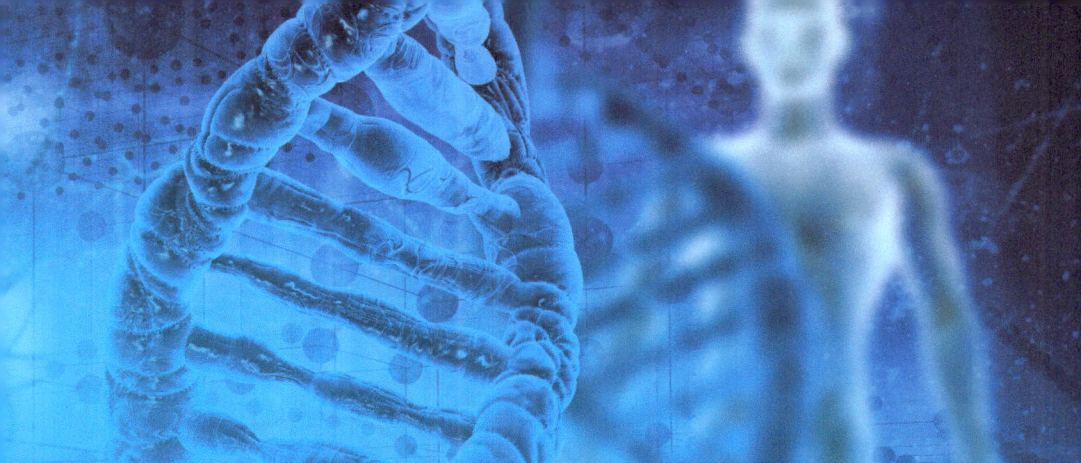

CHAPTER SEVEN
Genetic Engineering

Genetic engineering – the direct manipulation of DNA by humans – has existed since the 1970s. In 1972 American biochemist Paul Berg was the first person to create recombinant DNA (DNA molecules made up of genetic material from multiple sources) by combining the DNA from two different viruses.

In 1974 Professor of Biology at MIT Rudolf Jaenish created the first transgenic animal, which was a mouse that had foreign DNA introduced to its genome. Jaenish inserted virus DNA into an early-stage mouse embryo and found that this genetic change became present in every one of the mouse's cells but it did not pass the DNA changes on to its progeny.

However, in 1981 a collaboration between the University of Pennsylvania and the University of Washington developed a technique for genetic engineering that successfully passed down genetic changes to subsequent generations.

Genetically engineered mice are now widely used in medical research as models for human disease because their tissues and organs are similar to that of a human and they carry almost all the same genes.

The most common type of genetically modified mouse is the 'knock out' mouse, which has the activity of one or more genes removed. These have been used in medical research to study obesity, heart disease, diabetes, arthritis, substance abuse, anxiety, aging and Parkinson's disease. Specially engineered oncomice, that have had their tumor suppressing genes knocked out, are increasingly being used as models for human cancer.

GloFish

In 1999 a team of researchers from the National University of Singapore inserted a jellyfish gene that naturally produces a green fluorescent protein into the embryo of a zebrafish.

This resulted in the creation of a fish that was brightly fluorescent under both natural white light and ultraviolet light. Initially the fluorescent fish were developed for use as a pollution detector. The alterations in the zebrafish's genes gave it the ability to fluoresce as a bio-indicator and this genetic ability was used to detect pollution and chemicals.

Following this, the team created red fluorescent zebrafish by introducing a gene from sea coral and further colours have since been achieved by using genes from other sources.

The fluorescent zebrafish were patented and trademarked as GloFish by Yorktown Technologies before being sold commercially in the United States in late 2003 as the world's first genetically modified pet.

Glowing pigs

In 2006 National Taiwan University's Department of Animal Science and Technology used the same green fluorescent protein to genetically engineer green pigs. Researchers claimed that unlike previous attempts by scientists to create glowing pigs these ones were green inside and out.

Even their heart and internal organs were green. These pigs were used to research human disease because, as the pigs' genetic material encodes a protein that shows up as green, it was easy to spot. This way, scientists could track the stem cells injected into another animal easily without the need for a biopsy or invasive testing.

Mini Bama pigs

In 2015, researchers at the Beijing Genomics Institute (BGI) used the genetic engineering technique TALEN (Transcription Activator-Like Effector Nuclease) to create a miniature pig that was about the size of a medium-sized dog.

After considerable interest in the mini 'Bama' pig it has been reported that they are to be sold commercially in the future as pets. Each micro-pig will be sold for about 10,000 yuan (around $1,500) and customers will be able to select their animal's colour and coat pattern, which the BGI says it can achieve by manipulating the animal's genetic make-up.

Ethics

The most common form of genetic engineering used up until recently involved inserting new genetic material at an unspecified location in the host genome. In 2012 a new technique was discovered that is revolutionizing the field of genetic engineering, making it much cheaper, easier and more accurate to use.

"The ethics of the use of gene editing for altering traits in pets should be the subject of public debate," said reproductive biologist Willard Eyestone from Virginia State University in an interview with the Guardian in October 2015. "We must bear in mind that we have been altering the genetic make-up of pets for millennia, using the comparatively imprecise method of selective breeding, which sometimes results in less than healthy traits for the animal. In principle, gene editing should offer a far more predictable and humane alternative to selective breeding for all domestic animals."

Enter CRISPR-Cas9

Geneticist and biochemist Dr Jennifer Doudna from UC Berkley co-invented the CRISPR-Cas9 genome editing technique. It's a way to alter any organism's DNA, much like a computer user can edit a word in a text document.

Dr Doudna said,
"I feel very optimistic that within ten years in human health applications we're going to see the CRISPR-Cas9 technology used to repair mutations that are well known to cause genetic disease."

The CRISPR-Cas9 system was discovered when scientists studied how bacteria fight viral infections.

When a virus infects a bacterial cell, it injects its DNA into the cell, taking it over and killing it. Bacteria have evolved different ways to defend themselves against viruses. One of these is called CRISPR – Clusters of Regularly Interspaced Short Palindromic Repeats.

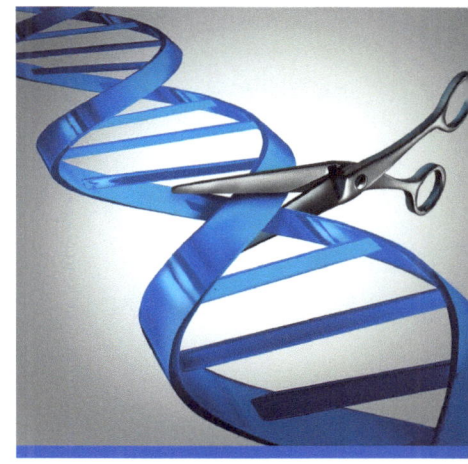

Bacteria take the pieces of viral DNA left in the cell and store them in the genome. This then helps bacteria to recognise those viruses should they try to infect the cell again.

Once the viral DNA has been inserted into the bacterial chromosome, the cell makes an exact replication of the viral DNA called RNA. RNA is a chemical cousin of DNA and allows interaction with DNA molecules that have a matching sequence.

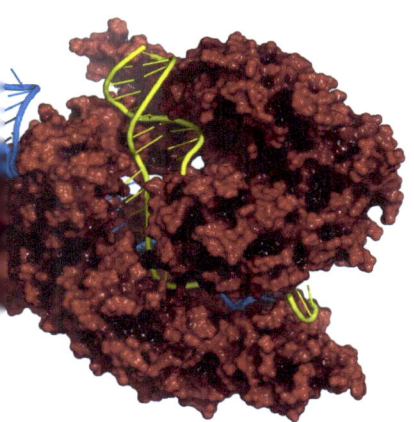

CRISPR-Cas9 in action

RNA binds to a protein in the cell called Cas9 and forms a complex that acts like a sentinel. It searches through the DNA in the cell to find sites that match the sequences in the bound RNAs. When these sites are found the Cas9 cuts the viral DNA very precisely through the double helix like a pair of scissors.

Cells naturally have the ability to detect broken DNA and to repair it. They do this either by reattaching the two ends of DNA or inserting a new piece of DNA at the site of the break.

In understanding how Cas9 works, Dr Doudna realised it could be harnessed as a genetic engineering technology to delete or insert specific bits of DNA into cells with incredible precision.

Dr Doudna describes previous genetic engineering methods as being, "similar to rewiring your computer every time you want to run a new piece of software" but the CRISPR-cas9 system she describes as "software for the genome" as you can simply cut and paste like you would with a word processing program.

The CRISPR-Cas9 system has been used as a research tool and has so far produced more muscular beagles, repainted the wings on butterflies, created a mushroom that doesn't brown when you cut it, removed malaria from mosquitoes and edited 62 genes in pig cells at once to make their organs suitable for human transplant. It is also being looked at as a way of bringing back extinct species – the future as foretold by Michael Crichton in his book Jurassic Park

Bringing back the dead

The last of the woolly mammoths (Mammuthus primigenius) died out nearly 4,000 years ago but one scientist is committed to trying to bring them back. George Church, a CRISPR pioneer at Harvard Medical School in Boston, Massachusetts, plans to use the genetic engineering tool to transform endangered Indian elephants into woolly mammoths so that they would be more resilient to the cold and could be released into a reserve in Siberia

Church recognises that this ambitious project is a huge undertaking yet one that is entirely possible with the CRISPR-Cas9 system. However, he says that it would be unethical to implant gene-edited embryos into endangered elephants so is looking at ways to build an artificial womb to fulfil the role of mother.

The CRISPR-Cas9 system works with any cell so it is a system with seemingly endless applications, which is why it has begun to cause some ethical concern among the scientific community. Just because we can do all these incredible things, does that mean that we should and where do we draw the line?

The creation of enhanced humans and des gner babies are no longer science fiction concepts and in the near future could become reality. Dr Doudna believes that scientists should proceed with caution.

CRISPR-Cas9 and humans

The first human trial involving the CRISPR-Cas9 system is already underway in China. It is being used to disable a gene called PD-1 in immune cells, which has a "switch" that many cancers evolve the ability to "turn off" so they can thrive. Cells are being taken from cancer patients and edited to remove the "switch" before being injected back into their body.

A trial in the US is due to start soon, which will take this process a step further. As well as disabling the PD-1 gene and two other genes researchers will add an extra gene engineered to make the immune cells specifically target tumors. It is hoped that by combing these two techniques it will make the treatment far more effective.

CHAPTER EIGHT
Gene Therapy

Gene therapy is the technique of using genes to prevent or treat disease. It is still very much in its infancy as a medical field but has recently provided some very encouraging results. It is hoped that in the future patients will be treated by having genes inserted into their cells to make changes to their bodies instead of using drugs or surgery.

Gene therapy works by replacing a mutated gene with a healthy copy, inactivating or 'knocking out' a mutated gene that is not functioning properly or introducing a brand new gene to the body that will fight disease.

How does it work?

It is very difficult to introduce new genes to a cell and keep them functioning so this has been the focus of much of the gene therapy research.

Nature has developed a very effective way of delivering foreign genes into the cells of living things – the virus. Scientists have discovered that this natural process can be harnessed as a vector to deliver gene therapy into the body for therapeutic purposes in the same way.

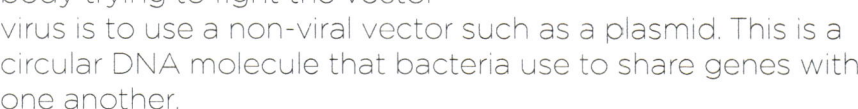

The advantages of using viruses in this way is that they are already good at targeting and entering cells – some can even target specific cells – and they can be modified so that they can't replicate or destroy cells.

Another method of delivery that is less effective but avoids the body trying to fight the vector virus is to use a non-viral vector such as a plasmid. This is a circular DNA molecule that bacteria use to share genes with one another.

The delivery of gene therapy can either be done in vivo (injected directly into the patient) or ex vivo (inserted into cells that have been taken from the body and once activated are reinserted). In some cases, gene therapy will only work if a gene is delivered to several million cells in a tissue and even if this is successful the new gene must be activated when it reaches its new location. Cells have a habit of shutting down genes that are too active or behaving unusually so researchers need to ensure that, once activated, a gene stays 'switched on'.

Gene therapy and the immune system

One challenge faced by scientists is that of the body's own

defense system. The immune system acts as a sentinel to identify anything foreign trying to enter the body and, as gene therapy usually uses viruses as a vector to carry new genes, getting them past the immune system can be tricky.

The consequences of an unwelcome immune response can be fatal and it was this that gave gene therapy research its greatest setback. In 1999, 18-year-old Jesse Gelsinger took part in a clinical trial to see if gene therapy could rectify the rare liver disorder that he had been born with. During the trial carried out by the University of Pennsylvania Jesse was injected with a corrected gene within a virus.

Unfortunately the viral vector triggered a massive immune response, which led to multiple organ failure and ultimately Jesse's death. As a result, the FDA suspended a total of 27 other clinical trials involving gene therapy over safety concerns while investigations were carried out.

To get around the risk of immune system response, researchers now either deliver viruses to cells outside the body and then re-introduce them or give the patient immune suppressing drugs prior to the gene therapy to temporarily disable their immune system.

Location, location, location

Targeting is also very important when it comes to inserting genes into someone's DNA, as it could be detrimental to their health if the wrong gene ends up in the wrong place. In order for a new gene to become a permanent part of a target cell it needs to 'stitch' itself into the DNA, but if it does this in an inappropriate part of the code it can actually disrupt the function of another gene.

This is exactly what happened following two clinical trials that ran between 1999 and 2006 involving children with the genetic condition Severe Combined Immune Deficiency (SCID). People with SCID have virtually no immune protection against bacteria and viruses, and can only escape illness by living in a completely germ-free environment.

Research into SCID revealed that a defective gene in the immune system (known as gamma c) was causing the condition. This particular variation of SCID is the most commonly found and represents 46% of all SCID cases.

By restoring the function of this important gene, researchers seemingly fixed the immune systems in the children who received treatment. However, five children later developed leukemia and it was discovered that the restored gene had 'stitched' itself onto a gene that normally helps regulate cell division. In disrupting the function of this gene the cells began to divide rapidly causing the blood cancer. Although four of the children were successfully treated using chemotherapy, one died.

Now researchers are using vectors that are able to specifically target 'safe' places in the DNA to integrate a new gene. Also, if genes are introduced to cells ex vivo researchers have the opportunity to check where they are located before returning the cells back to the patient.

Scientists at Milan's San Raffaele Telethon Institute for Gene Therapy reported that they had cured a group of children of ADA-SCID – the second most common form of SCID – in a trial spanning a decade. Instead of the gamma c gene being responsible, sufferers of ADA-SCID are affected by a lack of an essential enzyme called adenosine deaminase (ADA).

Eighteen children with ADA-SCID were treated with stem cell gene therapy between 2000 and 2010. Researchers removed bone marrow from the affected children under general anesthetic, added a working copy of the ADA-making gene to the cells and replaced them via an intravenous infusion.

Reconstruction of the immune system was observed from six months following treatment and was sustained over time with the presence of cells containing the healthy ADA gene being confirmed. All 18 patients were alive after a median follow-up of 6.9 years (ranging from 2.3 to 13.4 years).

The treatment, now called Strimvelis, took 14 years to develop. It is owned by GlaxoSmithKline and was approved for use in Europe in May 2016. However, gene therapy is not cheap and the $665,000 price tag could prove to be prohibitive.

Gene therapy restores sight

In 2001, the University of Pennsylvania and Cornell University College of Veterinary Medicine used gene therapy to give sight to three puppies born blind due to the genetic condition LAC (Leber Congenital Amaurosis).

Researchers identified a mutant gene that causes blindness and replaced it with a gene they had created in the laboratory.

The new gene was planted inside a viral vector and delivered by injection directly into the puppies' eyes. The virus 'infected' the eyes with the replacement gene and within a month the dogs could see for the first time in their lives.

The eye has proved to be a good area to receive gene therapy as the retina is easy to access and is partially protected from the immune system as viruses cannot move from the eye to other parts of the body.

Due to the success of this trial the procedure is now being used to give sight to children who have been born with the same rare genetic disease, which affects one in 80,000 people.

Blood disease success

Gene therapy has also been used to treat the genetic blood

disease beta-Thalassemia, a condition in which a person doesn't have enough red blood cells to carry oxygen around their body. Patients with the disorder are reliant on blood transfusions for their survival.

In 2007 a patient with severe beta-Thalassemia had blood stem cells (cells that continually make fresh blood cells to replace the ones that expire naturally) extracted from his bone marrow and had a working copy of the beta-globin gene inserted into them.

When the blood stem cells were returned to the patient's body they began to produce healthy red blood cells without the defective gene. After seven years the patients still had no need for blood transfusions.

In 2016, a team of researchers from the University of North Carolina published the findings from its study into dogs with FV11 deficiency – a genetic bleeding disorder where there is inadequate production of a blood-clotting protein.

Using a single gene therapy injection with the missing gene enclosed inside the common cold virus, the condition was corrected. As blood, kidney and liver function tests all showed that the therapy was safe with no unwanted immune response, the next step will be to conduct clinical trials in humans with the condition.

A cure for diabetes?

In 2013, researchers at the Universitat Autonama de Barcelona's Centre for Animal Biotechnology and Gene Therapy, cured a group of diabetic dogs.

Five dogs with Type 1 diabetes were given two different genes to give the body the glucose-regulating function it lacked. The two genes worked together to restore the body's ability to detect high blood sugar levels and then produce insulin to promote the uptake of blood glucose into the cells.

After a single gene therapy session the dogs were able to effectively control their own blood glucose levels for the four years that they were monitored. The monitoring also showed that the gene therapy was better at controlling the levels than daily insulin injections.

The success of this research project is a promising sign that the treatment could one-day work for humans suffering from the disease, which currently affects around three million Americans.

Gene therapy for pets

We are still in the very early stages of finding out what gene therapy is capable of and, most importantly, how results can be achieved safely.

In theory gene therapy could provide a cure for a whole host of genetic disorders found in pets but with the costs involved currently being seen as prohibitive to human patients it could be a very long time before we see it routinely being offered by veterinarians.

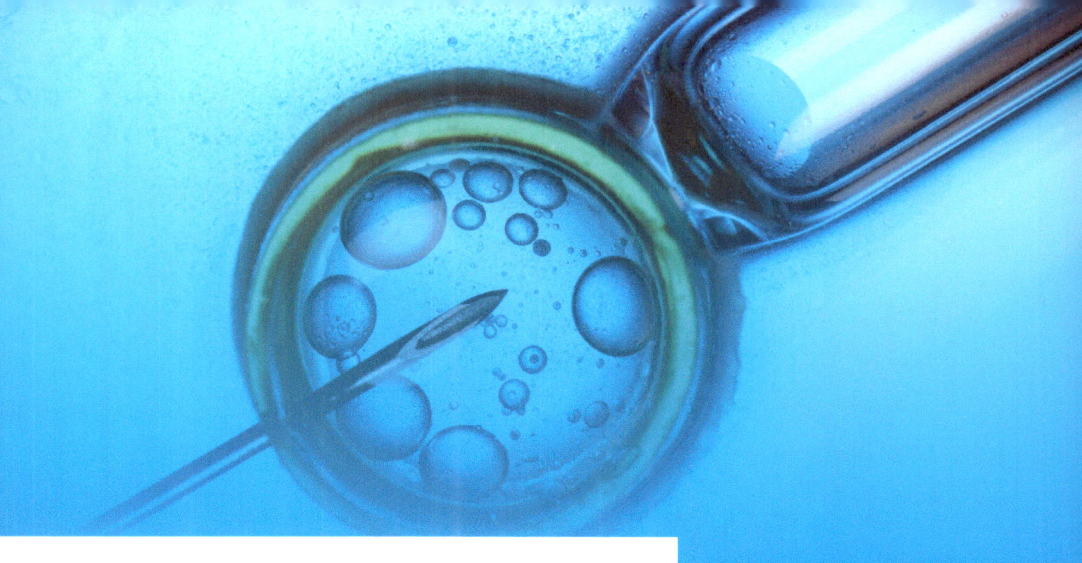

CHAPTER NINE
Cloning

Cloning is the process of producing genetically identical copies of an organism. This occurs naturally with some plants and single-celled organisms such as bacteria being able to reproduce asexually.

It is also how identical twins are created in humans and other mammals when a fertilized egg splits in two, creating two or more embryos that carry almost identical DNA.

In 1885 Hans Adolf Edward Dreish discovered that every cell in an early embryo has the genetic instructions needed to grow into a full organism. He demonstrated this by shaking a two-celled sea urchin embryo until the cells separated then observing how each cell grew into a complete sea urchin.

With this knowledge, research has been carried out over the past 50 years to find out how the cloning process works and ways in which man can control it through artificial cloning.

As the nucleus in a cell contains all the genetic information required to make an organism, researchers found that they could extract this information by removing the nucleus from an adult body cell.

Experiments have since been carried out using this extracted DNA and inserting it into enucleated eggs ('blank' eggs that have had their own nuclei removed) to effectively 'grow' genetically identical embryos.

Once the genetic information from an adult body cell has been inserted into a blank egg it requires an electric current from an electro cell manipulator to kick-start the cell division.

Once an embryo has begun to develop it can be implanted into the womb of a surrogate female animal to carry to term. This technique is called Somatic Cell Nuclear Transfer (somatic meaning any cell of a living organism that is not concerned with reproduction).

Dolly the sheep

In 1996 a landmark experiment led to the first mammal being created by Somatic Cell Nuclear Transplant (SCNT). A team of researchers from the Roslin Institute, University of Edinburgh created the famous sheep, Dolly, by transferring the nucleus from an adult sheep's mammary cell into an enucleated egg. After 277 attempts, only one egg produced an embryo that was carried to term by a surrogate mother.

The success of Dolly's arrival was particularly notable due to the fact that the cells used were adult ones. While every cell's nucleus holds a complete set of the genetic instructions, adult cells that are being used for specific functions tend to shut down the genes they are not using. When an adult cell nucleus is used it needs to be 'reset' to an embryonic state, as embryonic cells are ready to activate any genes they hold.

A year after Dolly was born the same team cloned another sheep, this time introducing a human gene that aids blood clotting into the sheep's skin cells. The resulting clone – Polly – produced milk that contained the human factor nine protein, which highlighted for the first time the potential medical and commercial uses for cloning in being able to engineer therapeutic milk.

Therapeutic cloning

In 2013 another groundbreaking cloning experiment was carried out by Shoukhrat Mitalipor and colleagues from the Center for Embryonic Cell and Gene Therapy at the Oregon Health and Science University in Portland. This was the first time that SCNT had been used to create a human embryo that could be a source of embryonic stem cells.

Embryonic stem cells are master cells that can grow into any type of cell found in the body and are naturally genetically matched to the donor. This means that they can be engineered to become any cell required and can replace damaged cells in a patient.

Therapeutic cloning or stem cell therapy is being hailed as a possible means to cure dozens of degenerative diseases from heart disease to Parkinson's disease. With just a little chemical encouragement an embryonic stem cell can become a new heart muscle, a neuron for someone suffering paralysis or a new pancreas cell needed to cure someone of diabetes.

In 2001 Great Britain became the first country to legalize the creation of human embryos for the purpose of harnessing stem cells for research. The condition of this use is that embryos are destroyed after 14 days and must not be used to create cloned babies.

There are concerns however regarding therapeutic cloning and the striking similarities between stem cells and cancer cells. Some experts say that the relationship between the two needs to be more clearly understood before the technique is used to treat human disease. Both stem cells and cancer cells have the ability to proliferate indefinitely and some studies show that after 60 cycles of cell division, stem cells can accumulate mutations that could lead to cancer.

First dog clone

The first successfully cloned dog, an Afghan Hound named Snuppy, was born in 2005. Snuppy was cloned by a team from Seoul National University whose leader Hwang Woo-suk went on to form his own cloning company – Sooam Biotech – and has since started to offer commercial pet cloning.

It was Sooam Biotech that produced the world's first commercial dog clone in 2011. The dog cloned was a Labrador called Lancelot owned by Edgar and Nina Otto from Florida.

The couple paid $155,000 for the pup clone, who they named Lancelot Encore, and they were very happy with the outcome saying that they bonded with Encore straight away as he looked and acted just like the original Lancelot.

Sooam Biotech has produced more than 700 clones of dogs for commercial customers and in 2015 announced it was teaming up with Chinese biotech company BoyaLife Group to open the Tianjin Animal Cloning Factory.

Once built it will be the largest cloning facility in the world and aims to produce up to a million cattle embryos a year to meet the demand for quality beef in China.

In 2015 Sooam Biotech also produced two clones of a deceased boxer dog for a British couple. This was a first for Britain and a first for Sooam Biotech as, when the cells were taken, the dog had already been dead 12 days.

The longest delay previously for successful cloning had been five days so this opened the possibility for increasing the timescale in which samples would need to be collected from a dog post-mortem.

For owners who are not in a position to be able to pay $100,000 for a clone of their pet, Sooam Biotech offers a cell storage service. The company will cryopreserve (preserved by cooling to very low temperatures using liquid nitrogen) a pet's cells for $3,000, which can be defrosted at any time in the future to make a clone.

In 2015, ViaGen – a Texas company that had been cloning horses and livestock – turned its attentions to commercial pet cloning. In October 2015 it produced two litters of cloned kittens and a Jack Russell terrier. ViaGen will clone your dog for $50,000 or cat for just $25,000 as it says cats are easier to clone. The company can also bank your pet's genetic material so you can choose to clone at a later date.

First cat clone

The first cat clone, named Carbon Copy or CC for short, was born in December 2001 at Texas A&M University. Genetically, CC was an exact copy of the original cat, Rainbow, just as if they were naturally occurring identical twins, yet they looked very different from each other.

Rainbow was orange mixed with patches of black, with a white belly and legs while her clone was a tabby. This is because the pattern of colours on multi-coloured animals is actually determined by events in the womb, rather than by genes. It is not known how long CC lived but she had a litter for four kittens in 2006 and reached at least 10 years old.

Cloning applications

Reproductive cloning has potential benefits in the fields of medicine and agriculture. By using clones in drug testing you can compare an animal's response to that of any number of genetically identical animals, meaning their responses should be uniform rather than variable as you would find in animals of differing genetic make-up.

In January 2008, the U.S. Food and Drug Administration (FDA) approved the use of cloning in agriculture after consultations with independent scientists concluded that meat and milk from cloned animals would be safe for human consumption.

Although the current costs of cloning make it unfeasible to use for routine meat and milk production, it does give researchers the green light to work on using cloning methods to produce animals with desirable traits, such as high milk yield or leaner meat.

Another application for cloning is to build up the population of endangered species. In 2003 an endangered Banteg ox and three African wildcats were successfully cloned but the results have received mixed reactions from conservation experts.

Some believe that cloning is going to be the saviour of those species we are set to lose, while others argue that populations of genetically identical individuals will lack the genetic diversity needed for a species to survive.

Cloning disadvantages

Reproductive cloning is actually a very inefficient technique – remember the team that produced Dolly the sheep did so on the 277th attempt. Some researchers have also observed health problems in mammals that have been cloned, such as increased birth size, organ defects, premature aging and problems with the immune system.

Dolly died when she was six years old, about half the lifespan you would expect from a sheep. It is thought that, because the cells used to create her were from a six-year-old sheep they were already aging.

> **"The central and continuing problem with cloning is aging,"** explains Dr Ron Bank, member of the Society for Veterinary Medical Ethics in an interview with Veterinary Practice News. **"A cloned animal is not a newborn, chromosomally speaking. It might look like a kitten but its body is already well along in life. As such, geriatric medical issues will come a lot sooner than usual. Why create for my pleasure an animal that will suffer old age maladies while middle-age joy is fleeting?"**

Another disadvantage, particularly if you are looking at pet clones is that, while they may be genetic copies they can never be completely identical because there is so much that is dictated outside of the genes.

An animals' character is influenced by early experiences and it's environment so it is unlikely that a clone will have exactly the same character and traits as the original.

A question of ethics

Since pet cloning and genetic preservation have become available commercially, albeit only for the rich, it has prompted debate within the veterinary profession about how ethical a practice it is.

Dr James A. Serpell, professor of animal ethics and welfare at the University of Pennsylvania told Veterinary Practice News,
"It's unclear at this point how many of the cloned offspring survive to term. It depends how far through gestation these embryos get and how sentient they are when they die."

Aside from the ethical concerns surrounding the embryos, there is also the welfare of the surrogate mothers to take into consideration. The implanting of embryos into the womb of a surrogate needs to be done under general anaesthesia.

Dr Serpell continues,
"I know it happens all the time with bitches who have trouble reproducing, but it's still an unnecessary medical procedure that the female dog has to endure in order to provide a surrogate placenta for this cloned embryo so that certainly raises ethical concerns. Should we be using dogs for that sort of purpose?"

ViaGen uses surrogate dogs from a large breeding partner and says it houses them in social groups within its biosecure facility where they receive daily interaction and play with staff.

After surrogacy the dogs are offered for adoption with the client, but if they decline then ViaGen will look at adoption elsewhere or using them again in the cloning program. ViaGen owns a feline surrogate colony for use when cloning cats and these are often put up for adoption afterwards.

It is unlikely that cloning will become a popular choice for pet owners in the future. Although a cat or dog cloned from a dearly beloved pet will have the identical genetic make-up as the original it may not look identical and personality-wise it could be very different.

Therefore, cloning should not be seen as a way to replace a departed pet. So much of an individual's character is a product of its experiences and environment in early life, which is impossible to replicate.

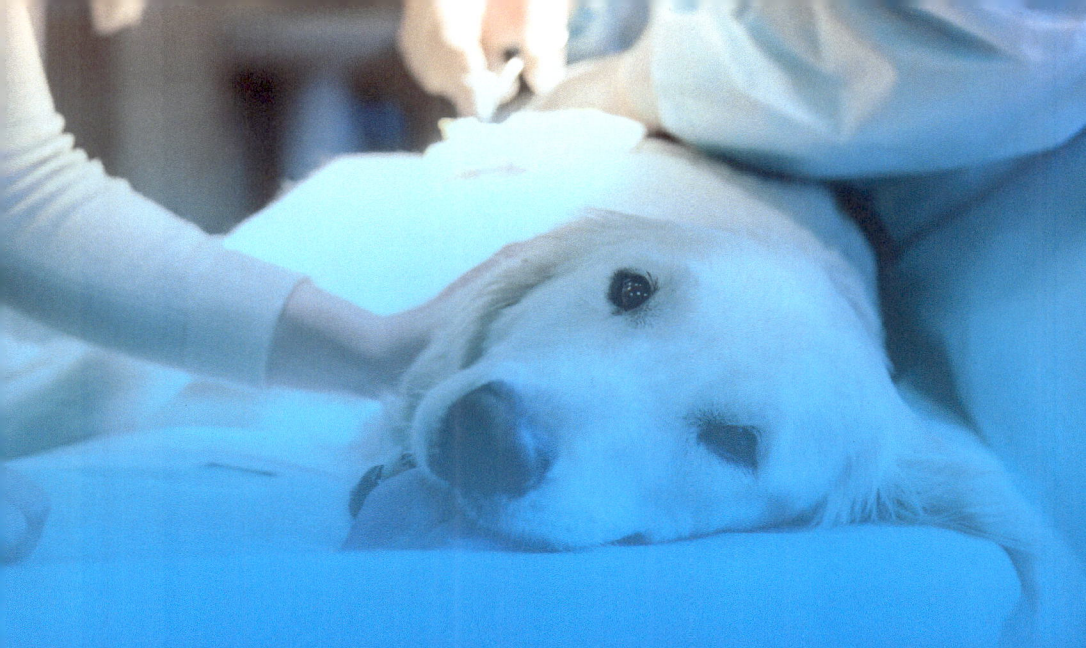

"I believe that every life has a beginning, a middle and an end. It's curious to me why a person would want to reanimate or re-create a specific life. The circle of life is an absolute, and helping people cope with this is more compelling to me than re-creating a specific life."

- Dr Susan Spence, member of the Society for Veterinary Medical Ethics in an interview with Veterinary Practice News.

REFERENCES

Chapter 1:

- Introduction to genetics
  http://www.nhs.uk/conditions/Genetics/Pages/Introduction.aspx
- National Human Genome Research Institute chromosomes fact sheet
  https://www.genome.gov/26524120/chromosomes-fact-sheet/
- What is a genome?
  http://www.yourgenome.org/facts/what-is-a-genome
- Genetics, DNA and heredity: the basics
  https://www.genome.gov/pages/education/modules/basicspresentation.pdf
- An introduction to DNA, genes, and chromosomes
  http://www.genetics.edu.au/Publications-and-Resources/Genetics-Fact-Sheets/FactSheetDNAGenesChromosomes

Chapter 2:

- How to sequence the human genome
  http://ed.ted.com/lessons/how-to-sequence-the-human-genome-mark-j-kiel
- Implications of the genome project for medical science
  https://www.genome.gov/25019925/online-education-kit-implications-of-the-genome-project-for-medical-science/
- Mapping and sequencing the human genome
  https://www.ncbi.nlm.nih.gov/books/NBK218244/
- Whole genome sequencing and its implications: can we know too much?
  http://www.medicaldaily.com/whole-genome-sequencing-and-its-implications-can-we-know-too-much-313308
- DNA research: this is the age of the genome. And there are exciting times ahead
  https://www.theguardian.com/commentisfree/2014/aug/03/genetic-research-britain-leads-the-way

- I had my whole genome sequenced — and so should you
  http://www.huffingtonpost.com/stephane-budel/i-had-my-whole-genome-sequenced_b_8288952.html
- Your full genome can be sequenced and analyzed for Just $1000
  http://www.popsci.com/cost-full-genome-sequencing-drops-to-1000
- The 100,000 genomes project
  https://www.genomicsengland.co.uk/the-100000-genomes-project/

Pet Genomes:

- Ostrander laboratory
  https://www.genome.gov/12513335/
- Dog genetic mapping
  https://www.youtube.com/watch?v=m0VJbciYpDU
- What is veterinary medical genetics?
  http://www.towncentercares.com/2015/11/what-is-veterinary-medical-genetics/
- Sequencing the cat genome
  http://genome.cshlp.org/site/press/CatGenomeSequence.xhtml
  http://www.nature.com/news/i-can-haz-genomes-cats-claw-their-way-into-genetics-1.16708
- 99 Lives
  http://felinegenetics.missouri.edu/99lives
- Comparative genomics
  https://www.genome.gov/11509542/comparative-genomics-fact-sheet/
  https://www.youtube.com/watch?v=3oNNozuY3dg
- Narcolepsy discovery
  https://vanwinkles.com/doctor-emmanuel-mignot-on-his-sleep-habits

Chapter 3:

- The hidden benefits of DNA testing for dogs
  http://www.petmd.com/dog/general-health/hidden-bene-

- fits-dna-testing-dogs
- DNA testing
  http://thebark.com/content/dna-testing
- Dog DNA tests: why your mutt's makeup matters
  http://pets.webmd.com/dogs/features/dog-dna-testing
- Dog DNA testing gets its day
  https://www.bloomberg.com/news/articles/2016-04-06/dog-dna-testing-gets-its-day
- Embark
  https://embarkvet.com/
- What you should know before you test your dog's DNA
  http://fortune.com/2015/06/25/dog-dna-tests/
- DNA-testing dog poo? You'd have to be barking
  https://www.theguardian.com/lifeandstyle/shortcuts/2015/apr/28/dna-testing-dog-poo-have-to-be-barking-council

Chapter 4:

- Solving the problem of genetic disorders in dogs
  http://www.instituteofcaninebiology.org/blog/solving-the-problem-of-genetic-disorders-in-dogs
- Top 5 genetic diseases of dogs
  http://www.cliniciansbrief.com/article/top-5-genetic-diseases-dogs
- Online mendelian inheritance in animals
  http://omia.angis.org.au/home/
- Inbreeding depression
  http://evolution.berkeley.edu/evolibrary/article/conservation_03
- It's all in the genes
  http://www.canismajor.com/dog/genetic1.html
- Genetic drift
  http://evolution.berkeley.edu/evolibrary/article/evo_24
- The pox of popular sires
  http://www.instituteofcaninebiology.org/blog/the-pox-of-popular-sires
- Illena and the seven sires
  http://www.dpca.org/BreedEd/index.php/component/content/article/92-illena-a-the-seven-sires

- The 6 most common genetic disorders in dogs
  http://www.petmd.com/dog/slideshows/6-most-common-genetic-disorders-dogs
- DNA testing and simple inherited disorders
  http://www.thekennelclub.org.uk/health/for-breeders/dna-testing-simple-inherited-disorders/

Chapter 5:

- Canine behavioral genetics: pointing out the phenotypes and herding up the genes
  https://www.ncbi.nlm.nih.gov/pmc/articles/PMC2253978/
- The behavioural genetics of dogs
  http://www.animalbehavioronline.com/dogbehavioralgenetics.html
- Holding back the genes: limitations of research into canine behavioural genetics
  https://cgejournal.biomedcentral.com/articles/10.1186/2052-6687-1-7
- How do dogs' genes affect their behavior? Your pet could help scientists find out
  https://www.washingtonpost.com/news/animalia/wp/2016/12/13/how-do-dogs-genes-affect-their-behavior-your-pooch-could-help-scientists-find-out/
- Genes underlying dogs' social ability revealed
  https://www.sciencedaily.com/releases/2016/09/160929092603.htm
- Secret of connection between dogs and humans could be genetic
  https://www.theguardian.com/science/2016/sep/29/secret-of-connection-between-dogs-and-humans-could-be-genetic
- Is our dogs' behavior genetic?
  http://www.whole-dog-journal.com/issues/19_9/features/Is-Our-Dogs-Behavior-Genetic_21514-1.html
- Darwin's Dogs
  https://darwinsdogs.org/?pg=about

Chapter 6:

- What's in a name? A lot when it comes to precision medicine
http://www.xconomy.com/national/2013/02/04/whats-in-a-name-a-lot-when-it-comes-to-precision-medicine/
- Precis on Medicine
https://en.wikipedia.org/wiki/Precision_medicine
- What's the Precision Medicine Initiative?
https://ghr.nlm.nih.gov/primer/precisionmedicine/initiative
- What's precision medicine?
https://ghr.nlm.nih.gov/primer/precisionmedicine/definition
- What is pharmacogenomics?
https://ghr.nlm.nih.gov/primer/genomicresearch/pharmacogenomics
- Precision medicine: an opportunity for a paradigm shift in veterinary medicine
https://www.ncbi.nlm.nih.gov/pmc/articles/PMC4996120/
- Going to the Dogs
http://genomemag.com/going-to-the-dogs/#.WKMNBH_KOvZ

Chapter 7:

- Gene-edited micro pigs
http://www.nature.com/news/gene-edited-micropigs-to-be-sold-as-pets-at-chinese-institute-1.18448
- Jennifer Doudna TED talk
https://www.youtube.com/watch?v=TdBAHexVYzc&sns=tw
- Things we can do with CRISPR-Cas9
http://www.popularmechanics.com/science/a19067/11-crazy-things-we-can-do-with-crispr-cas9/
- The early days of CRISPR
http://epigenie.com/the-early-days-of-crispr-with-dr-blake-weidenheft
- Realizing the potential of CRISPR
http://www.mckinsey.com/industries/pharmaceuticals-and-medical-products/our-insights/realizing-the-potential-of-crispr

- WTF is CRISPR?
  https://techcrunch.com/2016/11/06/wtf-is-crispr/
- First human patient treated using CRISPR
  http://www.popsci.com/crispr-tested-in-human-patient-for-first-time?con=TrueAnthem&dom=tw&src=SOC&utm_campaign=&utm_content=5882ffee19d6ba0008bae4cf&utm_medium=&utm_source=
- GloFish
  https://en.wikipedia.org/wiki/GloFish

Chapter 8:

- What is gene therapy
  https://ghr.nlm.nih.gov/primer/therapy/genetherapy
- Gene therapy tools of the trade
  http://learn.genetics.utah.edu/content/genetherapy/tools/
- Gene therapy for blindness
  https://www.technologyreview.com/s/543181/crispr-gene-editing-to-be-tested-on-people-by-2017-says-editas/
- Gene therapy cures deafness in mice
  http://www.globalfuturist.org/2017/02/new-gene-therapy-technique-cures-deafness-in-mice/
- Types of SCID
  http://www.adagen.com/types_of_SCID.html
- Strimvelis
  http://www.gsk.com/en-gb/media/press-releases/2016/gsk-fondazione-telethon-and-ospedale-san-raffaele-announce-publication-of-pivotal-safety-and-efficacy-of-gene-therapy-for-children-with-ada-scid/
- Everything you need to know about gene therapy's most promising year
  https://www.technologyreview.com/s/603206/everything-you-need-to-know-about-gene-therapys-most-promising-year/?utm_campaign=add_this&utm_source=twitter&utm_medium=post
- Hemophilia breakthrough
  https://www.hemophilia.org/Newsroom/Medical-News/Researchers-Make-Gene-Therapy-Breakthrough-in-Dogs-with-Factor-VII-Deficiency

Chapter 9:

- The history of cloning
  http://learn.genetics.utah.edu/content/cloning/clonezone/
- What is cloning?
  http://learn.genetics.utah.edu/content/cloning/whatiscloning/
- Cloning fact sheet
  https://www.genome.gov/25020028/cloning-fact-sheet/
- The future of pet cloning
  https://www.genome.gov/25020028/cloninghttp://www.veterinarypracticenews.com/the-future-of-pet-cloning/-fact-sheet/
- Sooam Biotech
  http://en.sooam.com/dogcn/sub01.html
- ViaGen
  https://viagenpets.com/
- First British couple to clone dead pet
  http://www.telegraph.co.uk/news/uknews/12133278/First-British-couple-to-clone-dead-pet-dog-pick-up-puppies-from-South-Korea.html

ABOUT THE AUTHOR

Dr. Gordon Roberts has had a lifelong love affair with the future of healthcare.

Growing up as a child in New Zealand, he would make regular trips to a nearby forest and, surrounded by some of the Southern Hemisphere's most striking landscapes, he would spend his time wondering what future health care would look like.

Even then, he had a strong feeling that what we consider to be an accepted standard in healthcare would be replaced by an entirely new paradigm. He saw a world where the lives of both humans and animals would be greatly improved and extended by the advances to come.

This forward-thinking start in life led him to where he is today, one of the world's leading futurist veterinarians with a passion for the medicine of tomorrow.

He is particularly interested in the intersection between conscious energy, medicine and technology.

Inspired by exponential advances in human medicine, Gordon is helping bring this knowledge to veterinary science and the broader pet welfare industry.

Gordon divides his time between his native New Zealand (where he lives with his wife, four children and many pets) and conferences, seminars and airport terminals around the world as he spreads the word about futurist vet opportunities, while dealing with his angel investments in future tech.

Want to read more about these exciting developments as they happen? All the latest news and discoveries from the future of veterinary medicine can be found on Gordon's website futuristvet.com

www.ingramcontent.com/pod-product-compliance
Lightning Source LLC
Chambersburg PA
CBHW041103180526
45172CB00001B/86